ANIMAL HOMES AND SOCIETIES

ANIMAL
HOMES
AND SOCIETIES

BILLY GOODMAN

Little, Brown and Company
Boston Toronto London

First Edition

ISBN 0-316-32018-8

Library of Congress Cataloging-in-Publication Data

Goodman, Billy.
    Animal homes and societies / Billy Goodman.   1st ed.
      p. cm. (A Planet Earth book)
    "A Tern Enterprise book" T.p. verso.
    Summary: Takes a look at how animals interact with each other and their environments.
    ISBN 0-316-32018-8
    1. Social behavior in animals Juvenile literature. 2. Animals Habitations Juvenile literature. 3. Territoriality (Zoology) Juvenile literature. [1. Animals Habits and behavior. 2. Animals Habitations. 3. Territoriality (Zoology)] I. Title. II. Series.
QL775.G66   1991
591.5'1   dc20   91-52590

A TERN ENTERPRISE BOOK

10 9 8 7 6 5 4 3 2 1

Animal Homes and Societies was prepared and produced by
Tern Enterprise Inc.
15 West 26th Street
New York, NY 10010

Editor: Stephen Williams
Designer: Judy Morgan
Photography Editor: Ede Rothaus

Additional photography: pp 6-7 © Mark Newman/Tom Stark & Associates; 8-9, 12-13 © Gerry Ellis/The Wildlife Collection; 38-39 © Ede Rothaus; 48-49, © D. Wilder/Tom Stack & Associates; 66-67 © Rod Planck/Tom Stack & Associates; © John Cancalosi/Tom Stack & Associates

Typeset by Classic Type, Inc.
Color separations by Excel Graphic Arts Co.
Printed and bound in Hong Kong by LeeFung-Asco Printers Ltd.

Dedication
To Andy and Denise

Acknowledgements
Thanks to Fred Singer for reading the entire manuscript and improving it in many places. For discussions about various species I thank Melanie Stiassny, Judith Winston, Dave Smith, John Harstead, Daryl Karns, John Terborgh, and Frank McKinney. My editor at Tern Enterprises, Stephen Williams, improved the manuscript with his comments. Finally, the staff of the library at the American Museum of Natural History was extremely helpful.

CONT

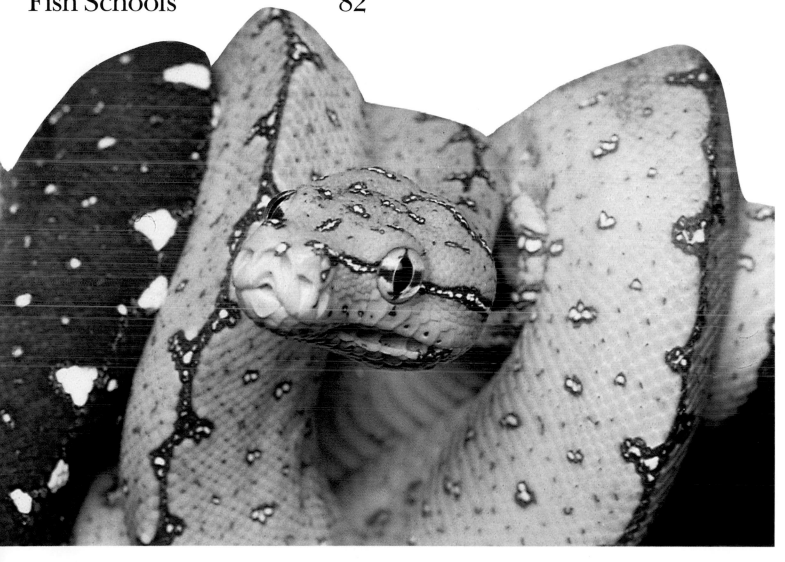

E N T S

INTRODUCTION

On pages 6 and 7, a bald eagle swoops in to join its mate at its bulky but sturdy nest. On this page (top) is a black noddy tern from Australia. Snakes, such as the Madagascar tree boa (bottom), tend to be relatively solitary creatures. Ground squirrels, such as the golden-mantled ground squirrel (opposite, left), in its burrow in Colorado, are often social creatures. They typically live in large underground colonies. A honeybee (opposite, right) gathers nectar on a milkweed flower. Hives may contain thousands of bees such as this forager, of which almost all are sisters.

© Gerry Ellis/The Wildlife Collection

© D.G. Barker/Tom Stack & Associates

At last count, just over one million *species,* or kinds of animals, had been discovered and named. Scientists suspect there may be ten million more animals yet to be discovered. Three-quarters of the known animals—and almost certainly as many of the still-undiscovered animals—are insects. Most of the rest are various kinds of *invertebrates,* or animals without backbones, such as sponges, worms, mollusks, spiders, and crustaceans. The kinds of animals most familiar to people, including fish, reptiles, amphibians, birds, and mammals, number fewer than forty thousand species. The variety of animal life on the planet is incredible. And all of these animals have their own ways of living, with their own societies and homes.

We usually think of a society as being a big group. But it doesn't have to be. Some animals only associate with other animals of their own species when they meet to mate from time to time—or perhaps only once in their lives. Other animals have a social system that resembles a

nuclear family: mother, father, and their offspring. Sometimes this grouping is expanded to include several adult females and males, and the many sets of offspring that remain with the family after they become adults. These offspring help their parents raise a new set of offspring. Still other animals live in large groups. Typically, an individual in a big group is not closely related to most of the other members of the group.

All animals also have homes of one sort or another. A home may be a structure that an animal builds or discovers and moves into, such as a bird nest, a beaver lodge, or a beehive. Or a home may be the general area in which an animal lives, called a home territory. The animal stays within this territory and keeps competitive animals out of it. A territory could be as small as the rock where a lizard lives, or as large as the fifteen square miles a mountain lion ranges over. Almost all animals have a home range. But not all animals build a specific home.

LIVING ALONE

Many species of animals live all or most of their lives alone. Sometimes the offspring will remain with the mother until they are old enough to venture out on their own. But the male and the female do not live together, they just meet when they need to mate. This type of society might seem very lonely to humans. But for some animals, living this way makes good sense.

Some animals would be spotted by predators much more easily if they were in a group. Sometimes animals that don't have any fear of predators still live alone. There are several possible reasons for this: Perhaps there isn't enough food for more than one animal, or for more than a female and her offspring, in the area, so a group of animals couldn't live there; or perhaps it's easier for the animal to capture prey by sneaking up on it, which is hard to do in a big group.

TIGERS

The largest species of cat is the tiger, which lives in Asia from India to Siberia. Tigers, especially the males, spend most of their lives alone. Adult males may measure 10 feet long and weigh 500 pounds. The largest tiger ever found by humans weighed 845 pounds. Tigers live in wooded areas that provide plenty of cover for the style of hunting they use, known as "stalk-and-ambush." They sneak up on their prey and leap on it or drag it down after a short chase.

One reason that tigers live and hunt alone is that a single tiger is less likely to scare away a prey animal, which may be faster than the tiger. Another reason for being solitary is that a single kill provides only enough meat for one animal. Even under the best conditions, tigers capture a meal only once or twice in ten tries. When a hunt is successful, a tiger will typically drag the carcass of the dead animal to a sheltered spot and gorge on it alone. It may eat more than fifty pounds of meat in one night and continue feeding on the kill for several days until the animal has been completely devoured.

A Bengal, or Indian, tiger emerges from cover. This carnivore fears no animal and has only one enemy: humans. There is only one species of tiger, but several subspecies are recognized.

Den Mothers

Female tigers usually have three or four cubs to a litter and raise them alone. The cubs are born blind and helpless in the *den,* a protected area in tall grass or in a depression in the ground. The male tiger doesn't help the female raise the cubs, but instead spends its time roaming far and wide, defending its territory and looking for food for itself. This arrangement is for the best, however, because a large male tiger wandering through the den could pose a danger to small, helpless cubs.

The cubs hide in the den while their mother hunts nearby. At about eight weeks of age the cubs begin following their mother outside the den. But they remain dependent on the mother's hunting ability for

Tigers have an excellent sense of smell (above). *Individuals can probably recognize each other by their odors. During the warm season, tigers are often found resting near water or standing in it to keep cool* (right).

their food until they are nearly two years old. Then they leave their mother's home range, looking for an area of their own.

Male and Female Territories

Both male and female tigers roam home territories that they defend against other tigers of the same sex. A male's territory is much larger than a female's, averaging thirty to forty square miles versus a female's eight to ten square miles.

The size of a tiger's territory depends on several factors. The tiger needs an area large enough to provide enough food to keep it alive. But it must accommodate other tigers in the region, too; when there are many tigers, each one will have a smaller territory.

By having a territory and getting to know it well, a tiger increases its chances of capturing food. In addition, males with ranges that surround several females' ranges have the chance to mate with all of them. Tigers mark the boundaries of their ranges in several ways. They spray urine on trees, bushes, and trails, and they use trees as scratching posts.

It was long thought that males and females avoided each other except during the mating season. But wildlife biologists have recently discovered that encounters between the sexes are more common than previously thought. Usually, a male that is patrolling its territory will spend a few hours with any female it comes across before continuing on its way.

© Nancy Adams/Tom Stack & Associates

ORANGUTANS

The orangutan lives in the dense rain forest of Borneo and Sumatra, in Southeast Asia. It is one of humankind's closest relatives in the animal kingdom, along with the chimpanzees and gorillas. The three "great apes," as they are called, are thought to have shared a common ancestor with humans about ten million years ago. Together, the apes, monkeys, and man form part of a group scientists call primates.

Orangutans are the most solitary of all the apes. Males live almost their entire lives alone. Females care for their offspring for many years, but otherwise also live alone.

Orangutans, such as this large male, are one of the few truly tree-dwelling apes. They live in the tropical rain forest of Southeast Asia and eat mostly fruit. Male orangutans are almost completely solitary. They accompany females only briefly for mating and occasionally encounter males at territory boundaries.

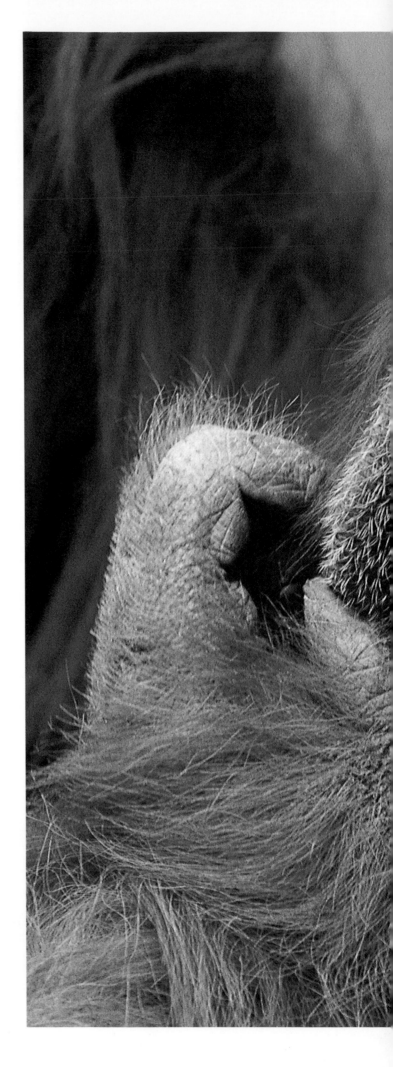

Female orangutans (above) *don't live with other adults, but care for a single offspring for as long as ten years. At right, a fifteen-week-old baby rides on its mother's back while she eats.*

At Home in the Trees

One reason that orangutans live alone is that they have no reason to live in groups. They are very big and strong and therefore don't worry about predators. They spend almost their entire lives in trees and can easily find enough food to eat by foraging for themselves. They are the biggest tree-dwelling animals in the world.

Orangutans use their feet as well as their hands to grip tree limbs as they swing through the trees. They tend to take their time when traveling, often stopping for hours in a single tree to eat fruit, which makes up the bulk of their diet. Orangutans are reported to be extremely smart creatures. They certainly do have a great knowledge of where to find fruit in their small territories. Sometimes they even find food by following birds to a fruit tree.

Orangutans usually sleep in trees, in nests of leaves and branches. Sometimes they make similar nests on the forest floor.

The Old Man of the Rain Forest

Female orangutans have their first offspring when they are about ten. They may have three or four offspring during their lives, almost always just one at a time. Young orangutans stay with their mothers longer than most animals. As the orangutan grows up, it becomes more and more independent and finally leaves its mother when it is about ten years old. Then the mother lives alone or has another orangutan. Orangutans live to be about thirty-five years old in the wild. They have earned the nickname, "Old Man of the Rain Forest."

Young orangutans have reddish coloring all over their bodies. Adults characteristically have a black face.

COBRAS

In the wild, cobras are mainly creatures of the night, resting in a hole or under rocks during the day. They live in the warm parts of India, where night temperatures don't get too low. They live alone because they can hunt better by themselves, and they can hide better from their predators when they're alone. Plus, they eat other snakes, which doesn't encourage a community to develop.

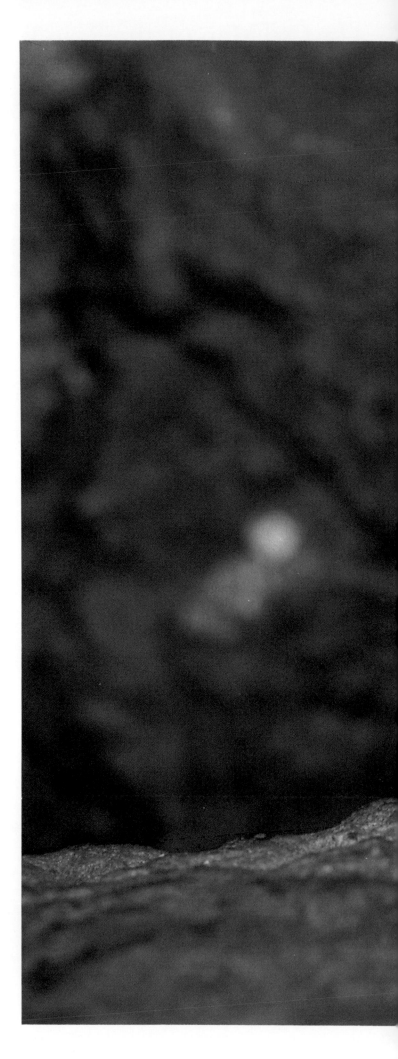

The king cobra is not the most venomous or aggressive snake known, but its venom can ill an elephant.

© Jeff Foott/Tom Stack & Associates

King cobras feed mainly on other snakes (above). Yet they are one of the few snakes to give any sort of care to their young. An Indian hooded cobra (opposite) extends its hood in a menacing pose.

Cobras Care for Their Eggs

Snakes in general don't give any care to their newly hatched offspring. Cobras, however, at least try to give the eggs a good chance of hatching. They build a more complex nest than any other snake and then defend the eggs.

When the female is ready to lay eggs, it uses its head and upper body to scrape together a pile of leaves, grass, and soil. The female cobra lays the eggs in a cavity on top and then covers the eggs with more leaves and remains on top of them until they hatch. This two-story nest may be three feet wide.

SNAKE CHARMERS

A snake charmer sits before a woven basket, blowing on a pipe and swaying slightly. Slowly, a hooded serpent rises from the basket. The cobra sways in time with the music. Is it magic?

Not at all. When the top is taken off a dark basket, a stunned cobra will lift its head in a normal defensive posture. Like all snakes, it can't hear a thing. But it sees well, and as the charmer sways his pipe the cobra will follow the pipe with its head, seeming to move with the music. If he's smart, the charmer sits just beyond the strike range of the snake. Even though this act isn't quite as dangerous as it looks, charmers are sometimes bitten and killed by their snakes.

BOWERBIRDS

On the island of New Guinea a bird with the tongue-twisting name of Vogelkop's gardener bowerbird lives apart from others of its species. But when it wants to mate, a male builds an incredible structure to attract a female. Called a bower, this structure looks man-made. In fact, when Westerners first saw these structures, they thought the native people had built tiny copies of their own huts for some unknown purpose.

The bower of this species looks like a domed tent. Males build them on the forest floor. First, the male clears an area of debris. Then it builds a pillar of twigs around the base of a small tree. Finally, it fashions a dome out of twigs using the pillar as a support.

This bower was built by a MacGregor's bowerbird in the rain forest of Papua New Guinea. It is virtually identical to the bowers built by Vogelkop's gardener bowerbirds. The bower, shaped like a May pole, is built around the base of a young tree out of small twigs. It can even have a door-like entrance and a "front lawn" that the bowerbird has cleared.

© S. Pruett-Jones/VIREO

© Gerry Ellis/The Wildlife Collection

Above: *This bower is called an "avenue" bower because the two walls of grass define a corridor. It is where courtship and mating take place.* Right: *This male bowerbird has placed colorful fruits and green shells in front of its bower in order to attract a female.*

Avenue Bowers

Other bowerbirds build what are called "avenue" bowers. They construct two walls of twigs, forming an avenue or alley. They'll decorate the entrance to the alley with berries, stones, bones, beetle shells, and even man-made material such as car keys or pen tops. Two species even paint the walls, using pigment made from fruit and their saliva.

Attracting a Mate

The dome is open on one side, giving it a cavelike entrance. The finished bower is about six feet in diameter and three feet high. Inside, the bower is carpeted with soft moss. The male bowerbird collects flowers, colored fruit, beetle shells, and even bits of colored plastic and places them in heaps in front of the nest.

All this, just to impress a mate!

Females, it seems, are attracted to these well-decorated bowers, and if they like what they see and if the male's dance is acceptable, they will mate with him. Then the female goes off to lay eggs and raise the offspring alone. The bower is never used again.

© Andy Day

A caddis fly larva's shelter protects its soft body from predators.

CADDIS FLIES

Some of the most remarkable animal homes are found underwater, built by caddis flies, insects that are some of the animal world's most amazing builders. Caddis flies are a group of more than five thousand species of insects that—despite the name—are not flies. They are members of an order called Trichoptera (which means "hairy winged" in Greek). The name refers to the adults, which look like moths. And like moths, they can often be found flying around electric lights at night.

Most of a caddis fly's life is spent alone underwater in the larval stage. The caterpillar-like larvae build the underwater shelters, which help protect their soft bodies from predators, such as dragonfly larvae or fish. Adults emerge from the water and live only a day or so until they mate and the female lays its eggs.

The larvae inhabit rivers, streams, lakes, and ponds worldwide and are easy to find. Examine rocks, logs, or plant stems for their well-made homes. The tiny structures are like tents made of evergreen needles or grass stems cemented together. In swift streams, the shelters are made of pebbles and grains of sand. The largest may measure one inch in length. The larvae produce a waterproof silk from their bodies that they use to cement their homes together and attach them to rocks or logs. Many of the larvae also spin a net, which they hold like a fishing net in the current to catch little bits of food floating by.

CLOSE FAMILIES

The most important reason for living in a nuclear family is that two parents are then available to help raise the offspring. Two parents may be better than one in gathering food for the young and in protecting them against predators.

With birds, for example, one parent usually has to sit on the eggs, keeping them warm so that they will develop properly. The other parent then hunts for itself, its mate, and the young. With other birds, the young eat so much that sometimes both parents are kept busy full-time catching insects or gathering fruit.

The offspring usually leave the family group when they are mature enough to mate and have offspring themselves. In this way, they avoid mating with relatives, which is called *inbreeding* and can lead to unhealthy offspring.

GIBBONS AND TITI MONKEYS

Most of the over two hundred species of *primates*—the order of mammals that includes humans—are sociable creatures. (The Orangutan is an exception, see pages 18 to 21.) Some monkeys and apes have social systems that resemble nuclear families.

Gibbons, which are small apes of Southeast Asia, and titi monkeys of South America live in nuclear families. In both groups (there are nine species of gibbons and three species of titi monkeys), an adult will form a bond with a member of the opposite sex that often lasts for life.

There are several advantages to family living for these animals. They live in territories, and living in a group makes it easier to defend a territory. Also, they are small (gibbons rarely weigh more than fifteen pounds as adults; titi monkeys weigh only two pounds or so), and they would be in greater danger from predators if they were alone than they are in a group.

A large gibbon bares its teeth. This tree-dwelling ape swings through the tree tops by its arms, a form of locomotion called "brachiation."

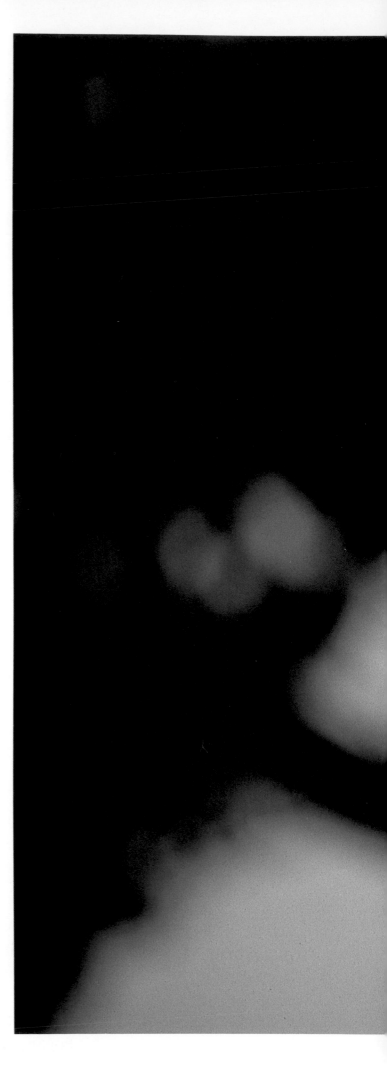

© Brian Parker/Tom Stack & Associates

Gibbon babies are weaned around one year of age or slightly later. They are then cared for largely by their fathers until they can move about on their own at about age three.

Gibbon Groups

A typical gibbon group consists of four animals: an adult male and female (the parents), and two off-spring. Females give birth to one off-spring every two or three years. Juveniles remain with their family until they are seven or eight years old, before leaving the family and looking for a mate and a territory of their own. The older juveniles play with the infants and sometimes help defend the territory.

Gibbons spend almost all their time high in the trees of dense rain forests. Scientists have observed that gibbons spend one-third of their waking hours eating. They eat fruit primarily, but they also eat young leaves and insects. A large part of their day is spent traveling from one feeding or resting spot to another.

Gibbons are among the few species of apes that don't build a nest. They sleep in the dense tops of trees near the center of their territory. They may use the same tree every night for weeks.

Titi Monkeys' Group Defense

While gibbons defend fairly large territories—they may be almost as large as one square mile—the smaller titi monkeys defend smaller territories. The titis' territory averages less than fifty yards in diameter, in the middle of dense rain forests. They scream and jump about when the territorial line has been crossed by an intruder. But usually when one family meets another at a territorial border there is a lot of brou-haha but no real fighting. In this way, titi monkeys are like common songbirds, which also make a variety of calls when they detect an intruder in their territory, but rarely end up fighting. The noise and commotion are alternatives to actually hurting each other.

Family groups consist of the parents, which often mate for life, and usually one or two offspring. A young titi monkey will leave its family group at about three years of age to find a mate and establish its own group.

Fathers play the major role in caring for the offspring in this species. They cover the infant when it's raining and stay close during storms to prevent the infant from falling out of the tree. One scientist found that more than 90 percent of the time, fathers are the ones who carry the offspring. In addition, by sharing food with the young monkey, the father teaches it how to eat solid food. Of course, until the young monkey can eat solid food, it drinks only its mother's milk.

The whole family sleeps huddled together on a large branch. Often, all their tails are intertwined as they sleep, a behavior that is unique to titi monkeys. Before going to sleep, they spend a lot of time grooming each other, removing small parasitic insects from each other's fur. The father usually spends the most time grooming the smallest infant.

A young titi monkey rides around in the trees on the back of an adult, probably its father. They can move about on their own at about four or five months.

DUNG BEETLES

There are thousands of species of dung beetles, and their family arrangements are about as close as insects get to having a nuclear family. The beetles are one of the few types of insects where both the male and the female care for the offspring. These insects, also called scarab beetles, are found all over the world, except in polar regions. The largest type of dung beetle can grow larger than a small bird. They eat the *dung*, or feces, that other animals leave on the ground. Dung is mostly just undigested food, and so it is nutritious. By feeding on dung and by burying it, dung beetles help fertilize the soil.

Most dung beetles are *nocturnal*, meaning they sleep during the day. They stay in during the day to avoid being eaten by birds and other predators that hunt in the daylight hours. At night they go out and look for dung. There is a lot of competition among beetles for the dung, and each has to work fast to get a piece.

A Brazilian dung beetle rolls a ball of dung as big as itself. It will eventually bury the dung and eat it or use it to provide food for its young.

Males and Females Work Together

While different dung beetles have different ways of gathering and burying the dung, one common species makes balls of dung and rolls them away from the dung pile before burying them. Males and females work together to make the ball of dung, roll it to a nice spot, and bury it. Depending on the size of the beetle, the dung ball may be as small as a pea or as large as a tennis ball. To roll it, a beetle typically stands on its front legs and pushes the ball with its rear legs. Often, the male pushes while the female rides on top.

When they've reached a good spot, possibly thirty feet away, the male buries the ball with the female on it and goes underground himself. The pair then feeds on the ball and mates.

Beetles go through four stages from the beginning of their lives to the end. After two adults mate, the female lays eggs on the surface of the dung. An egg will hatch into a soft-bodied, wormlike *larva*, also called a *grub*. The grub becomes a *pupa*, which resembles the adult, but is softer. The pupa turns into the adult beetle. This transformation is fueled by the nourishing dung where the eggs were first laid.

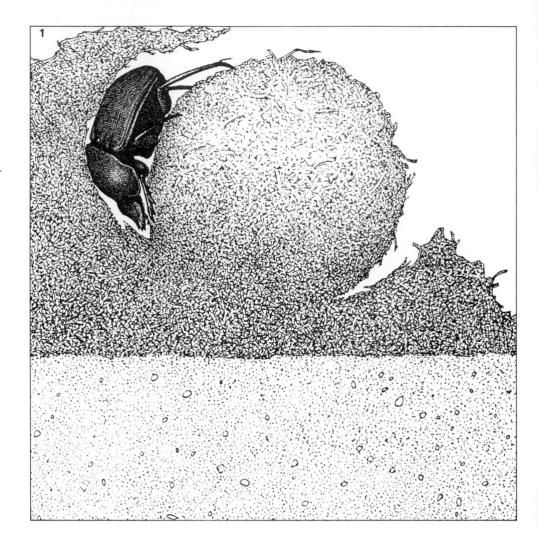

Here, the routine of the female beetle of the species Kheper aegyptiorum *is illustrated. First the beetle cuts the ball from the dung and shapes it (illustration 1). Then she rolls it towards a burial place (illustration 2). Once she finds a good spot she burrows into the ground to bury the ball (illustration 3). Finally, she lays a single egg in the dung ball, and the ball serves as food for the larva (illustration 4).*

NORTH AMERICAN BEAVER

No animal besides humans can match the North American beaver in its ability to make major, long-lasting changes in the way the earth looks. Just the sound of running water is enough to stimulate a beaver to start its dam-building behavior. Across North America, many wooded valleys with streams have been turned into ponds and marshes by beaver dams. Beavers are at home in the water, but they waddle around awkwardly on land and are easy targets for predators. So, if they can, beavers turn a fast-moving stream into a wide and sluggish body of water, where they can safely swim to their food sources without exposing themselves. Very large beaver dams might be as long as a football field, but that length is unusual. Usually they are much smaller. Beavers build dams to enlarge an existing pond or to create a new one, where they can build their home, called a lodge.

Beavers start the dam by shoving large logs into the stream bottom. They may brace the logs against underwater rocks or against trees standing at the edge of the water. They continue to add sticks and mud until the dam holds back the water. The dams are well made. Some are sturdy enough to ride a horse across. They usually last for generations of beaver families.

Beavers are most at home in the water. Much of their behavior is focused on creating a large, still body of water that is their "neighborhood."

At Home in the Beaver Lodge

In the newly created pond, where they will be safe from predators, beavers build their lodges. Beavers prefer to cut trees for the lodge right at the edge of the water, because they feel so vulnerable to predators on land. They may build a lodge into the bank, tunneling up through the mud. Or they may build it in deeper water, by piling up water-logged branches until they've created an artificial island.

The lodge has several entries, all of them underwater, so that the bea-ver must dive down to enter the lodge and then swim back up out of the water once it's inside. This keeps most intruders out. Mud caked around the outside of the lodge provides some insulation. But the beavers leave part of the roof mud-free, to allow air to move in and out of the chamber.

Only one beaver family occupies a lodge. The family consists of the parents, who usually pair for life, and one or more litters of kits. Each year, one to five kits are born. The kits nurse for about six weeks after birth. Then all the other family members, especially their father,

bring solid food to them.

The kits are able to swim within hours of their birth. But because they float too well, they can't dive down through the exit passage to leave the lodge. They will stay with their parents for at least one year, and sometimes more than two years.

A Woody Diet

By flooding a valley, beavers never have to venture far from water to find the foods they eat, especially aspen and willow trees. In the spring and summer they eat tree leaves and grasses. In the fall and winter they eat woodier material, such as stems and branches. They even store a cache of tree branches inside their lodge so they won't have to go out as frequently in the dead of winter.

Territorial Animals

Beaver families hold a territory, which they mark with a stinky substance called *castoreum* from their scent glands. They often mark small mounds of material dredged from the pond so other beavers will know whose territory it is. Marking is most intense in spring, when young beavers are leaving their families and looking to start families of their own. When an intruder crosses into a beaver's territory, the beaver will sometimes slap the water with its tail to warn the other beavers in the lodge.

The demand for beaver pelts is one of the reasons North America was explored in the 1600s and 1700s. Beavers were almost wiped out at the time, but have since rebounded.

Beavers in Alaska's Donali National Park pull branches from the water (opposite, top). In the summer beavers eat many leaves; in winter they eat woody plants such as aspen or willow (above). A beaver plugs a leak in its dam (opposite, below).

BIRDS OF PREY

Birds of prey, such as ospreys, eagles, and falcons, often mate for life, raising generation after generation of offspring. While birds in general are well known for their nests, birds of prey—also known as raptors—are noted for building huge, clumsy-looking nests.

Golden eagles, for example, nest on high cliffs. Made of branches as long as six feet, their nests are awkward looking, but sturdy. They're often reused year after year, with the nesting pair adding to them each season. One nest is reported to have weighed two tons. Bald eagles also make enormous nests, some larger than a king-size bed. Remarkably, these nests are usually in the tops of trees.

One of the reasons eagles and other raptors often reuse nests is that good sites—big cliff ledges or large trees with sturdy upper branches—are rare. Once built, the nests last a long time.

Eagle nests are at or near the tops of trees large enough to support their great weight. Notice the baby bald eagle beneath its parent.

LIFE PARTNERS

These animals mate for many years, or even life:

Adélie penguins	Eagles	Mute swans
Albatrosses	Foxes	Parrots
Badgers	Geese	Petrels
Barn owls	Gibbons	Ravens
Beavers	Hooded seals	Sea gulls
Coyotes	Marmots	Terns
	Mongooses	

Here, clockwise from above, are a red-tailed hawk, which is a common species found in woods and open country in the United States; a prairie falcon guarding its chicks at its nest in the northwest United States; and Ferruginous hawk juveniles at their nest in Colorado. This species is common on the Great Plains of the United States.

Division of Labor

Raptors tend to divide the labor of raising chicks. The male hunts and provides food for its mate, which sits on the eggs to keep them warm and help them *incubate* (develop in the shell) faster. When the eggs hatch, the male continues to hunt for the whole family, while its mate watches over the chicks.

When the chicks get big enough, the mother leaves them for periods in order to hunt. At first the mother tears up the meat she and the father catch so the young can eat it. But soon the mother gives them bigger pieces and lets them tear it up. This helps teach the chicks how to hunt. Eventually the birds learn to fly and are able to hunt on their own.

© Wendy Shattil/Robert Rozinski/Tom Stack & Associates

PEREGRINE FALCONS MAKE A COMEBACK

Peregrine falcons are graceful, fast-flying birds that prey on other birds. They divebomb flying targets as large as ducks, hitting them at speeds of one hundred miles per hour or more and knocking them out of the sky. Peregrines have been known to catch their prey even before it hits the ground.

Peregrine falcons were almost wiped out in the eastern United States (but not the western states) by the 1960s, due to the pesticide DDT, which damaged their eggshells. The poison was intended to keep crops free of insects. But the birds that peregrine falcons ate were also contaminated with DDT. The DDT caused the peregrine falcon eggshells to become thinner. Most eggs cracked before the chicks could hatch, and the peregrine population dropped alarmingly.

Now that DDT has been banned in the United States, peregrine falcons are making a comeback. They have nested historically on remote cliffs near water, and some of their traditional nest sites are being reused. But peregrine falcons have also begun nesting in some cities, where skyscrapers serve as artificial cliffs and pigeons make good prey. Minneapolis, Baltimore, Boston, and New York are among the cities that have had nesting peregrine falcons.

EXTENDED FAMILIES

In some animal species, the offspring remain with their parents after they are sexually mature. Sometimes the mature offspring only stay with their family an extra year or two. In other cases, offspring may remain with their parents as long as they live.

The social groups that result when mature offspring stay in the group are larger than nuclear families. One advantage to this is that older offspring may help their parents raise another set of offspring. Another advantage is that the larger group may be better able to capture prey, as is the case with lions and African wild dogs. Or the additional family members may help the group defend itself against predators, as is the case with Florida scrub jays and elephants.

LIONS

Lions are found in Africa, in the countries below the Sahara desert. They prefer to live in grassland areas, but they can also be found in wooded areas and along the edge of the desert. Because of their size and strength, they aren't threatened much by predators (with the exception of man), and they don't build shelters; they sleep instead in shady, protected areas, often in tall grass. Lion groups, called *prides*, often have over a dozen members.

A pride controls a home territory that varies in size, depending on how much prey is available. It may be as small as eight square miles if lots of food is available, or as large as one hundred and fifty square miles if prey is sparse. As other cats do, lions mark their range with urine and will chase away lions that don't belong to their pride.

Though people often call it the "King of Beasts," the lion starts life in a humble way. At birth it weighs about three pounds, less than most human babies and less than most of the young of the various species it will someday eat. It is blind and helpless. The lioness, or female lion, keeps it and its littermates well hidden.

After about six weeks, the lioness introduces the cubs to the rest of the pride. An average pride has between four and twelve lionesses in the group. This often makes for many cubs in a pride, when more than one female gives birth at about the same time. The typical pride has two to six adult males.

The lionesses in a pride tend to be related to each other, as sisters or cousins. The adult males may also be related to each other. Sometimes they are brothers. But the adult males are not related to the adult females.

This is a typical pattern with many animals that live in extended family groups. If all the offspring stayed in the group they were born into, chances are that brothers would mate with sisters. To avoid this inbreeding, one or both sexes leaves its family when it reaches adulthood.

Fighting Males

With lion families, the males leave and the females remain in the pride. After leaving the pride of their birth, some of the males may travel together, looking for another pride. If they find one populated by weak males, they will kick them out. Sometimes when males take over a pride, they kill any young cubs living there. This violent and seemingly bizarre behavior makes biological sense to the killer males who want their own offspring. They kill off the cubs so that the females won't spend years nursing another male lion's cubs. By killing cubs that are not his own, a male lion can cause their mother to become ready to mate with him sooner than would otherwise happen.

Cubs drink their mother's milk for most of their first year. They may also drink the milk of other mothers in the pride. Cubs start following their mothers to kills and competing with other animals in the pride for meat when they are about one year old. This is how they learn to hunt and defend themselves.

A female lion and cub rest in the shade in East Africa (right). In Kenya, a male lion roars (below). Female lions make most of the kills for the pride. Opposite, a female watches over the carcass of a gnu.

Female Hunters

Almost certainly the meat that the pride eats will have been killed by the pride's lionesses. Lions may run as fast as thirty-five miles per hour over short distances, but their prey is often even faster, which is why lions depend on stealth when searching for food. Several females usually hunt together, greatly increasing the chances that they will make a kill. The hunters spread out and stalk the prey slowly. One or more will circle behind the prey to try to drive it back toward the advancing group.

The males are less effective hunters. But when females kill prey, the larger males push the females out of the way and eat first. And a full-grown male can put away a lot of meat; a pair of males has been known to eat as much as ninety-five pounds in twenty hours.

Well-fed lions will spend most of their time—up to twenty hours a day—sleeping.

ELEPHANTS

While lions usually spend sixteen to twenty hours a day sleeping or resting, elephants must spend that much time eating. The difference is due to the kind of food each animal eats. Lions eat huge amounts of high-protein, high-calorie food, and one kill may last several days, during which time the pride needn't move. But elephants eat poor-quality plant food, only about half of which is digestible. An adult may have to eat more than three hundred pounds of plants a day—mostly grass—to maintain its health and weight. Adults continue to grow throughout their lives.

Like lions, elephants live in extended family groups, called *herds*. But unlike lions, adult male and female elephants don't live together. The males live alone or in small herds. There is no evidence of social bonding in the male herds. Instead, the male groups appear to be chance gatherings that break up as the males move about the habitat looking for food or for females to mate with.

The females live in herds of several adults and their calves. The adult females are usually related. They may be sisters or, because elephants live so long, they may be mothers and their grown daughters. The oldest, largest elephant, the matriarch, is the leader of the group.

Elephants eat a huge amount of plant material every day. Here,

an adult female and some younger animals munch grass under

a tree.

A baby elephant (right) is dependent on its mother for many years, but will also be cared for by other adults or older offspring in the group. The typical elephant group consists of several adult females and their offspring of both sexes (opposite, top). Elephants use their trunks to suck up water and squirt it into their mouths (opposite, below).

© Gerry Ellis/The Wildlife Collection

Finding Protection in the Herd

Elephant calves are dependent on their mothers for a long time. They nurse for about four years, and they don't reach adulthood until age eleven or twelve. A calf can nurse from any mother in the group, which is a great advantage of living in a herd with several adult females. Older female offspring will help look after the younger ones, prodding calves to keep them with the group, and making sure that they come to no harm. Female calves remain in the herd after they reach adulthood. However, the males begin to act aggressively at this age, and they are kicked out.

Adult elephants are subject to no predators other than humans (who kill elephants to collect the ivory from their tusks, and pose a serious threat to the survival of elephants in the wild). But young elephants may fall victim to lions, hyenas, or tigers. If a group of elephants feels threatened, the adults face the threat and the calves hide behind them. The matriarch takes the lead, standing tall in the center and ready to charge.

Wandering Without a Home

Elephants don't build homes or nests. They just move around looking for food and water. If the weather is hot, they will rest under the shade of a tree to avoid overheating. Elephants roam over a large home range to find food, water, and shade. A study of one African elephant herd revealed an average home range size of three hundred square miles—about five times the size of Washington, D.C. Elephants sleep only a few hours a night, wherever they happen to be.

FLORIDA SCRUB JAYS

Scrub jays have what biologists call "helpers" at their nests to help the parents take care of their offspring. A scrub jay pair that has helpers usually has one or two, though some have as many as six. The helpers are often mature offspring of the adults they're helping.

Most birds leave their parents soon after they can get food for themselves. Usually a one-year-old bird is sexually mature and can try to find a mate and have chicks. Scrub jays are also able to mate at age one, but many postpone breeding to help their parents or other relatives.

Florida scrub jays are one of many species of birds that have helpers at the nest to assist in raising the young.

HELPERS AMONG OTHER SPECIES

Helpers are found in many species of animals, including more than sixty species of birds. Here are the names of some of the animals.

Chimpanzees	Groove-billed anises	Lemurs
Eider ducks	Laughing kookaburras	Monkeys
Elephants		Porpoises

Scrub jays (above and right) have helpers, usually older off-spring, that help their parents until they have a chance to breed on their own.

Guarding the Nest

A scrub jay helper's main role is keeping an eye out for predators and chasing them away or giving an alarm call. About 80 percent of lost eggs or killed nestlings are due to predators, especially snakes. Helpers prevent a lot of deaths: Twice as many offspring survive with helpers around as without.

Scrub jay helpers do not help the parents build the nest, which is a rough, open cup of twigs. The nest is usually about two to ten feet above the ground. Helpers begin to feed the young one to two days after they hatch. Male helpers tend to bring more food than females, and older helpers bring more than younger helpers.

Why Be a Helper?

What do the helpers get out of the arrangement? After all, they've had to postpone breeding in order to help other birds raise offspring.

If a young scrub jay did leave its parents as soon as it could forage on its own, it wouldn't have a very good chance of finding a breeding territory. Their territories are in sandy, open areas of Florida, with few trees and bushes. The trees tend to be full of birds, with not much space for new birds to set up territories. A male scrub jay that remains with its parents and helps them may one day be able to take over all or part of its parents' territory. Helpers are closely related to the offspring they help raise. Scrub jays don't help unrelated birds.

© Don and Esther Phillips/Tom Stack & Associates

HONEYBEES

Although a honeybee hive has the population of a medium-sized city—eighty thousand bees or more—it is really just one big family. The head of the family is the queen bee. During most of the year there is only one queen in a nest, and it is the only bee to lay eggs. Its daughters, the workers that make up the vast majority of the bees in the nest, are sterile (not able to reproduce).

A tight mass of honeybees crawls over the comb. A honeybee hive is dark, and the bees communicate with each other by sound, smell, and touch.

EXTENDED FAMILIES 63

A honeybee colony has just a single queen (top), unless it is about to split into two colonies. As this picture shows, the queen is larger than the workers. The lower left section of the photograph shows larvae in their cells and at the upper right are capped brood cells. Wild African honeybees form a tight swarm (right); the queen is probably in the center.

Females Do the Work

As the name suggests, the workers do most of the jobs essential to honeybee society. They feed the brood (baby bees), giving them honey and a substance they make themselves called "bees' milk." Later in their five-week lives the workers help build or repair honeycomb. During this time they have glands that make wax. Toward the end of their lives the wax glands dry up and the workers become foragers.

There are no "king" bees in a honeybee nest. Males are produced only at certain times of the year, and their only job is to mate with queens from other nests.

Worker bees are always female. Some of them gather sweet nectar from flowers in the field and bring it back to the hive, which is located in a hollow tree or other sheltered spot. Inside the nest workers crawl around in complete darkness, storing pollen and nectar, caring for the brood, or doing general housekeeping.

The Hive

Honeybees live in hives made of waxy plates, called combs, which hang vertically in the hollow tree. The bees make the combs in two layers, each layer containing hollow, six-sided cells. The cells open on each side of the comb. The cells are where pollen and nectar are stored and where eggs are laid.

The Language of the Honeybees

When a worker returns from a successful foraging trip, it will be loaded with the nectar or pollen

from one species of flower. Many new foragers will be hanging around the comb, waiting for information on where to find food. The successful forager tells them with a dance.

If the flowers are very close, the forager will simply walk in a circle on the vertical comb. The smell of the pollen or nectar the forager carries will be the clue used by other workers to find the flowers.

When the food source is farther away, the forager flies in patterns in the air to tell others where the food is. The path of the pattern is always roughly in the shape of an 8. The important information is contained in the line that forms the middle of the figure eight.

The direction of the straight line tells what direction the food is from the nest. If the forager points straight up the vertical comb, it means the food lies in the direction of the sun. If the forager points straight down, it means the food lies in the direction exactly opposite the sun. The forager waggles its rear end very fast, producing a buzzing sound. The longer it waggles, the farther the food is from the nest.

Several bees may be giving directions at one time in the hive. The waiting workers judge which food source is best by how enthusiastic the foragers' movements are. The patterns are repeated over and over. Eventually, workers leave the nest and fly to one of the food sources.

A forager collects pollen, storing it on its hind legs.

LARGE GROUPS

Some animals live in very large groups called *colonies, troops, schools,* and *flocks.* The animals in these large groups aren't all part of the same family, as honeybees are. They live together for other reasons.

Sometimes they live together because all of the best nesting sites are in one location.

Animals also may live in a large group so that there are more eyes to watch for predators.

One drawback to living in a large group is that there has to be a lot of food in the area to support the population. The more individuals there are, the more mouths there are to feed. Some animals that normally live in larger groups form smaller groups when food is scarce.

OLIVE BABOONS

Baboons are extremely social African monkeys that spend much of their time on the ground. They live in large troops that sometimes number more than one hundred. The troop members stay together all the time.

Baboons sleep together in trees or on the ledges of cliffs. They never make nests, rather they sleep directly on the branches or rocks. Closely related baboons often sleep near each other. They also travel together. Females with small infants walk in the center of the troop, surrounded by the troop's highest ranking males. Other adult males and females march in front and behind. And they eat together, consuming a wide variety of plant foods, ranging from grass to the flowers of acacia trees. When they are able to kill a small animal, they will also eat meat.

In Gombe preserve in Tanzania, where Jane Goodall has studied chimpanzees, an olive baboon adult and youngster sit together.

Olive baboons eat many kinds of foods, from plant material to, occasionally, meat. Here, an adult sits on a rock eating a baobab fruit.

© Stan Osolinski/Marvin Dembinsky Photo Associates

Mothers provide most of the care to olive baboon young, but males, not necessarily the fathers, also help.

Groups Within Groups

Female baboons tend to stay with the troop they're born into—as do lions and elephants. The result is that there are many groups of relatives—called kin groups—within a single troop. There are usually about twice as many adult females in a troop as adult males. Adult males come to the troop from outside groups and are not closely related to the females.

Baboon males and females do not form pairs. Instead, they form what scientists call "friendships." A female will tend to have one or two male friends with which it spends much of its time. Each male in turn may have several female friends. When females are ready to mate, they will mate with several males, but prefer to mate with their friends. Males help their friends and provide some care to their friends' offspring—which may or may not be of their own blood.

Grooming is widespread among baboons. The groomer will use its teeth and hands to rake through another's hair, picking out, and often eating, insects, bits of dead skin, and clumps of matted hair.

Battling Males

The adult males that enter an olive baboon troop are not usually closely related to each other. But in addition to forming lasting relationships with females in the troop, males will form temporary relationships with other adult males.

When a female is in mating condition, a male will typically guard it and try to be the only male to mate with it. Another male may have a hard time stealing the breeding female from the guarding male. So two males sometimes team up to drive off the guarding male. The male that asked for help usually gets the female. The helper is performing what is called an *altruistic* act, because he risks being injured in a fight to help the other male. But later, the helper may ask for help. And it's likely to get it. This trading off of help is called *reciprocal altruism*.

PORTUGUESE MAN-OF-WAR

The Portuguese man-of-war is a floating animal that is found throughout the warm oceans of the world. It looks like a large jellyfish. But it's not really just one animal. Rather, it is a colony of hundreds of members that are attached to one another. The attached animals are called *zooids*. Each has a special function to carry out, much as the different organs of other animals have different functions. None of the animals could live without the others.

The uppermost zooid is a gas-filled float. The float keeps the Portuguese man-of-war suspended from the ocean surface, and can be about ten inches across. It sometimes serves as a sail, catching the wind and moving the colony around the ocean. Hanging from the float are the other zooids that make up the man-of-war. Stringy tentacles hang down as far as thirty feet. When stinging zooids on these tentacles touch fish they release a poison that kills the prey. Then tube-shaped zooids digest the food for the entire man-of-war.

Other zooids serve to reproduce the man-of-war, with eggs and sperm. A Portuguese man-of-war gets its start in life the same way humans do—from a single fertilized egg. But different zooids "bud off" the growing larva to form partly independent structures.

The gas-filled float of a Portuguese man-of-war looks like a blob of jelly. But it is only one part of a larger, more complex animal.

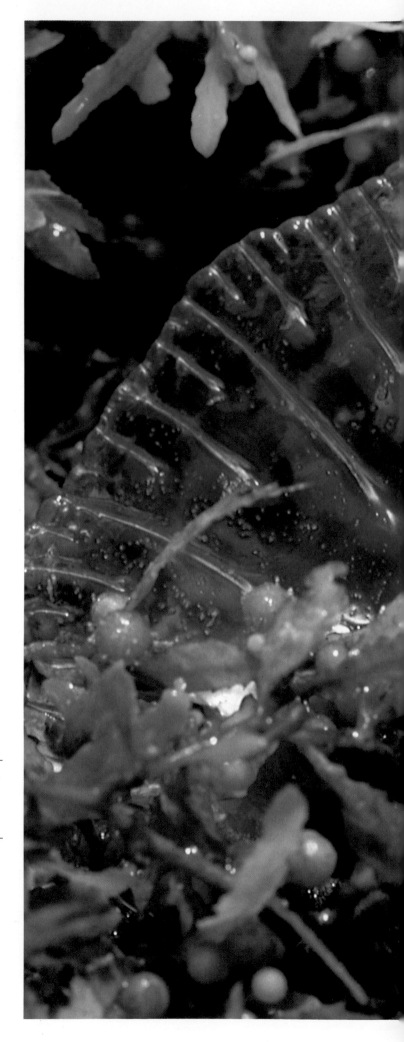

BATS

One-fourth of all species of mammals are bats, and bats are the only mammals that can fly. Bats live in a variety of social systems, from small groups in hollow trees to huge colonies in caves. In the southwestern United States, cave colonies of Mexican free-tailed bats may contain fifty million individuals. Most bats spend the day sleeping in their *roost,* or home, hanging upside down. Just before sunset, bats start to move around, and then at dusk they fly out of their roost to feed. Some bats eat insects. Others eat fruits and other foods. Some bats can eat half their weight in food each night.

Typical roost sites are caves, tree holes, or tree tops. Bats spend about half their lives in their roost sites, so they choose them carefully. Species that like a fairly constant temperature often roost in caves.

In places that have cold winters, bats may *hibernate* (pass the winter with little activity) in caves. Hibernating bats are often found with their wings pulled tightly around them, hanging upside down in huge clusters. This behavior helps them hold in heat and also slows the loss of moisture from evaporation.

Many bats roost in trees. Among the most interesting are tent-making bats. These tropical bats bite through the ribs of large leaves so that the leaves bend over, forming a protective umbrella. The bats then hang by their feet from the holes they've made with their bites.

Mexican free-tailed bats emerge at dusk from the entrance to

Carlsbad Caverns, New Mexico.

ECHOLOCATION

In the late 1700s, an Italian scientist named Lazzaro Spallanzani discovered something amazing about bats. A bat with its eyes covered could easily fly around a dark room filled with obstacles without hitting them. But a bat with earplugs could not. It seemed as if bats were somehow using sound to navigate, but Spallanzani couldn't hear the sounds they were using.

A little more than fifty years ago, an American scientist named Donald Griffin learned why. The bats were indeed using sound, but at too high a pitch for humans to hear. Griffin used a special microphone to pick up the high-pitched clicks. He found that the bats emitted high-pitched sounds which, like radar, bounced back to them. They could read the echo like a three-dimensional map.

This form of navigation is called echolocation. Most, but not all bat species have the capacity to do this. Other animals also use echolocation, including oilbirds, seals, whales, and dolphins.

The grey-headed flying foxes (above), hanging in a characteristic pose, are one species of a group of about 200 species of flying foxes that live in the tropics and feed mainly on fruit. They are also called megabats because they tend to be large. Gambian epauleted bats, also a type of flying fox, hang together (right). A vampire bat bares its teeth (opposite), showing why it has a bad reputation.

Bat Families

Bats don't usually bond in pairs. Instead, adults mate with several individuals. Often male and female bats will live in different roosts. Usually there is a mating season that lasts for only a few weeks, in the spring, fall, or winter, depending on the type of bat. Sometimes female bats will leave their colony to join a special nursery colony when they become pregnant.

Most types of bats have only one offspring a year. Since bats don't build nests, the offspring hold onto their mothers as they hang in their roosts. Female bats nurse their young for six to eight weeks.

Scary Creatures

Bats scare many people. This is probably due to two factors: the reputation of the vampire bat, a species that feeds on blood, especially from cattle; and the weird appearance of many bats, with their small eyes, huge ears, and strange growths around the nose.

There are nearly one thousand species of bats, and one entire group of nearly two hundred species is known as the flying foxes because their faces resemble fox faces, with small ears and large eyes. They often roost (sleep or rest) in huge groups in trees. They'll fold their wings over themselves for protection. Most flying foxes feed on fruit and nectar and don't use echolocation to find prey.

© Gary Milburn/Tom Stack & Associates

HUNTING MOTHS

Capturing a moth is a common task for a bat, and it illustrates just how complex bats are. The bat sends out several pulses of sound a second, and it can only listen for the echoes when it is not making its own squeaks and beeps. The echoes are some two thousand times quieter than the bat's own sonar pulses, but somehow bats pick them up. Bats can also tell the difference between a moth or another insect or a branch.

While looking for prey, a bat sends out an orderly stream of pulses. But when prey is detected, the bat increases the rate of pulses so that it can better track the insect.

Moths have evolved ways of knowing that a bat has detected them. Moths can hear bat sounds from about one hundred feet away, but bats can't detect moths until they are within ten feet. When moths hear prowling bats, they begin to fly directly away from the sound. If the bat has gotten close enough to detect the moth, flying away will not help the moth escape, unless the moth dives down to the safety of vegetation.

DOLPHINS

Even though they live in the ocean (and some rivers), dolphins are more closely related to humans and cows than to fish. But their stream-lined bodies are adapted to rocketing through the water.

Dolphins are typically found in groups of ten to one hundred, though schools of more than one thousand have been seen. With some species, such as the spinner dolphin of the Pacific Ocean, it's thought that subgroups of four to eight animals stay together for long periods, shifting in and out of the larger schools.

A group of common dolphins leap in the Sea of Cortez, off Baja

California. This species forms huge feeding schools.

Dolphins Help Other Dolphins

Because dolphins are so hard to study in the open oceans, very little is known about their social behavior. Still, they are known to be altruistic, in that they will help another dolphin while putting themselves in some danger. In one instance, a saddle-backed dolphin in the Mediterranean Sea was hit by a harpoon. Other dolphins in the school of about fifty came to help it. They pushed the wounded animal to the surface, where it could breathe. They did this three times and then all the dolphins, including the wounded animal, dived out of sight.

Intelligent Animals

Dolphins are extremely intelligent and social animals. One prominent scientist believes that dolphins probably fall somewhere between the dog and the rhesus monkey in intelligence.

Dolphins can cover a lot of territory in search of fish to eat. Often their prey lives in schools, so that when prey is discovered there will be lots of it. By spreading out, the dolphins can cover a wide area. They use echolocation, much as bats do, to find prey. When a school of fish is located, the dolphins alert each other.

A group of common dolphins seen from underwater (opposite). Above are Atlantic bottlenose dolphins at the Dolphin Research Center, in Florida.

BRYOZOANS

Bryozoans are tiny invertebrates found mostly in the ocean. Most species form colonies on a rock or piece of seaweed. The colonies may be a few inches across and look like a patch of moss or lichen.

Recently, Judith Winston, of the American Museum of Natural History, and some colleagues discovered many species whose entire colonies fit on sand grains. As you might expect, the animals are small—less than one-hundredth of an inch long. And of course the colonies are small, too.

The colony on a sand grain gets its start when a single young animal—a larva—settles onto the surface and attaches itself with a kind of glue it secretes. The single animal reproduces itself without mating. This process, called budding, results in a colony of genetically identical individuals.

Some individuals are specialized for feeding. They send out an organ that looks like a brush at the end of a tentacle to trap tiny particles floating by in the water. Other individuals are specialized for defense or reproduction.

FISH SCHOOLS

About two thousand species of ocean fish and an equal number of freshwater species live in schools that are tight groupings of hundreds or thousands of fish. Some schools may contain a million fish.

Even with so many fish grouped together, the movements of the individuals are remarkably well coordinated. The fish use their eyes to stay together and make the same movements at the same time. They also have a unique sense called the lateral-line system. The lateral line is made of hair cells that run along the sides of a fish. These cells can detect changes in pressure as water moves along the fish.

The members of a school tend to be about the same size, which makes staying together much easier. Small fish would have a hard time keeping up with larger ones.

Silversides form huge schools. Notice that the fish are close to the same size and maintain a relatively even spacing.

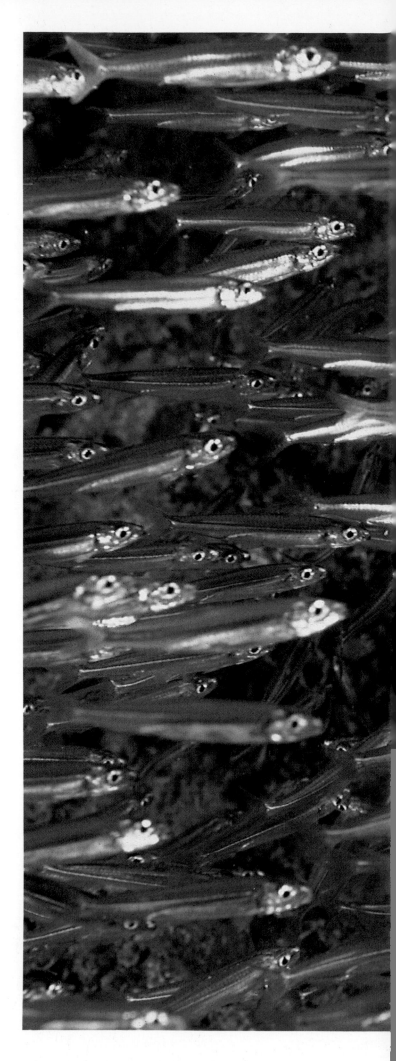

The Purpose of Schools

Schooling seems to have several purposes. The flashing movements of the school may confuse predators. And since the schooling fish stick together, it may be hard for a predator to pick out one to attack.

Something else is at work as well, what scientists call the "selfish herd" effect. In a school, the fish at the center are protected from predators by the layers of fish around them. The drive to be in the center may help keep the school together.

Schools of Predators

Some predators that hunt schooling fish also travel in schools. They may treat the school like one huge meal. One scientist has described a shark taking bites out of a school as if it were taking bites out of an apple.

© SharkSong/M. Kazmers/Marvin Dembinsky Photo Associates

Glass sweepers school near a wreck in the Red Sea (above). Grunt fish (opposite, top) and bigeyes (opposite, bottom) also form schools. The best explanation for many schools seems to be that they offer protection against predators.

BIRDS OF A FEATHER

Many birds, including sea gulls, terns, cliff swallows, bee eaters, and penguins, live at least part of the year in huge colonies. All of these birds leave their nests to get their food, and one advantage of nesting colonially is that birds can learn the location of food from other birds that return to the colony. A second advantage of nesting in a colony is that the group provides protection against predators. In one scientific experiment, larger colonies of cliff swallows detected snakes quicker than smaller colonies.

For example, the red-billed quelea, an African bird which feeds mainly on grass seeds, lives in some of the largest colonies known. One tree may hold one thousand nests, and one colony may contain ten million birds. When a small foraging flock finds food, it attracts others quickly. When some queleas leave the colony to get food, others follow them.

Gulls and terns often form large nesting colonies. These great crested terns are at a colony in Australia.

Cliff Nesters

Sometimes birds nest in colonies because the available space for nesting is limited. For example, black-legged kittiwakes, a kind of gull, nest colonially on cliffs near the sea. Good nest sites are uncommon, so the birds compete for them. As with many colonial nesters, kittiwakes are monogamous; they live as a pair within the colony. Most pairs stay together from year to year. Kittiwake nests are simple affairs: They make a small mud platform on a ledge to keep the eggs from rolling off.

Another type of bird, the cliff swallow, builds gourd-shaped nests of mud attached to cliff walls. The average colony in the central United States has about 325 nests; some colonies have as many as 3,000 nests.

© Diana L. Stratton/Tom Stack & Associates

© Diana L. Stratton/Tom Stack & Associates

Cliff swallows make mud nests (top *and* above) *that are closed on top (one of those* above *is incomplete or has been damaged). They feed on insects, captured in flight.*

Huddling Together for Warmth

Emperor penguins breed in colonies in Antarctica, the coldest habitat for any bird. At least part of their reason for breeding in large colonies is to stay warm. They mate in the Antarctic fall, which is in April or May. The females lay a single egg on the ground and then head out to sea to spend the winter looking for food. The males gather in tight groups within the colony—as many as ten birds in a space about as big as the hood of an average-sized car. Each male watches an egg, carrying the egg on its feet, tucked underneath the feathers of its belly. The males fast for two months or more, until their mates return at the end of the winter to relieve them. Another type of penguin, the Adélie penguin, which also nests in Antarctica, builds a small nest of rocks to keep its eggs above the water that comes when the snow melts in the spring.

Haystack in a Tree

The sociable weaver, a sparrow-sized bird of southwest Africa, builds a nest that may house hundreds of birds. A single nest may be several feet thick and thirty feet long.

Several birds work on the nest together, pushing stems into rough bark to make a solid mass, which they then hollow out. They work quickly, making a single chamber in a day so that the builders can have shelter.

When it is finished, the nest contains many chambers for different pairs of birds, with a common roof over all. The nest offers good protection from the cold desert winters. This allows the sociable weavers to breed during the winter, when their main predator—the Cape cobra—is inactive.

© Anna E. Zuckerman/Tom Stack & Associates

It may seem surprising to see penguins amid vegetation, as are these king penguins on South Georgia Island. But many species live on islands in the south Atlantic or Pacific oceans where snow and ice are not permanent. Some have difficulty staying cool in the heat.

THE EVIDENCE IS EVERYWHERE

   The variety of animals in the world is astounding. And each seems to have a unique way of living. Some spend almost their entire lives alone. Others seem to need to be surrounded by thousands, or even millions of their own kind. Each has a reason for the way that it lives. Evidence of animal homes and societies is all around us. Look towards the tops of buildings in cities and you're bound to see a flock of pigeons. Keep your eyes to the ground on a summer day and you'll probably see an ant colony carrying food back to its nest. Look around you and try to observe all the complex ways that animals live.

Adélie penguins flocking together on King George Island,

Antarctica.

© M. Timothy O'Keefe

METRIC CONVERSIONS

U.S. Units	Metric Equivalents
LINEAR MEASURE	
1 inch	2.54 centimeters
1 foot	0.3048 meters
1 yard	0.9144 meters
1 rod	5.0292 meters
1 mile	1,609.3 meters
1 furlong	201.168 meters
1 league	4.828 kilometers
AREA MEASURE	
1 square inch	6.4516 square centimeters
1 square foot	929.03 square centimeters
1 square yard	0.836 square meters
1 square rod	25.293 square meters
1 acre	0.405 hectares
1 square mile	2.5899 square kilometers
CUBIC MEASURE	
1 cubic inch	16.387 cubic centimeters
1 cubic foot	28.316 cubic centimeters
1 cubic yard	0.765 cubic meters
WEIGHT	
1 ounce	28.350 grams
1 pound	453.592 grams
100 pounds	45.3592 kilograms
1 ton	0.90718 metric tons

Temperature Conversions

°F	°C	°F	°C
32	0	125	51.7
38	3.3	130	54.4
42	5.6	135	57.2
46.4	8	140	60
50	10	145	62.8
55	12.8	150	65.6
60	15.6	155	68.3
65	18.3	160	71.1
70	21.1	165	73.9
75	23.9	170	76.7
80	26.7	175	79.4
85	29.4	180	82.2
90	32.2	185	85
95	35	190	87.8
100	37.8	195	90.6
105	40.6	200	93.3
110	43.3	205	96.1
115	46.1	210	98.9
120	48.9	212	100

To convert Fahrenheit degrees into Centigrade, subtract 32, multiply by 5 and divide by 9. To convert Centigrade into Fahrenheit, multiply by 9, divide by 5 and add 32.

INDEX